5

Len and Anne Frobisher

Heinemann

Heinemann is an imprint of Pearson Education Limited, a company incorporated in England and Wales, having its registered office at Edinburgh Gate, Harlow, Essex, CM20 2JE. Registered company number: 872828

www.heinemann.co.uk

Heinemann is a registered trademark of Pearson Educational Limited

ISBN 9780435208660 (Pupil Book)
ISBN 9780435208745 (Teacher's Version)

10 9
11

Illustrated by Andy Peters
Cover illustration by Andy Hunt
Typesetting and layout by J&L Composition, Filey, North Yorkshire

Printed in Malaysia, CTP-PJB

| UNIT | TOPIC | PAGE | UNIT | TOPIC | PAGE |

Contents
Autumn 1ˢᵗ half term

Spring 1ˢᵗ half term

Summer 1ˢᵗ half term

Number problems 1

1 In a competition Sue scored 6430 points and Greg 7010. Pam scored exactly half-way between Sue's and Greg's scores.

How many did Pam score?
6720

2 In every book there are 250 raffle tickets. Altogether ten complete books are sold.

How many raffle tickets are sold altogether? **2500**

3 On a Bank Holiday, 2176 planes take off from Heathrow and 1897 from Gatwick.

From which airport did more planes take off?
Heathrow

How many more? **279**

4 On a parade ground 8300 soldiers parade in groups of 10 rows with 10 soldiers in each row.

How many groups of soldiers are on parade? **83**

5 Lulu has three darts. She throws the first into single 9, the next into double 9 and the last into treble 9.

How many did Lulu score with her three darts? **54**

6 At a railway depot there are 20 trucks. The trucks are divided equally between two engines.

How many trucks does each engine have? **10**

7  I divide exactly by 100. My first two digits are equal. The sum of my four digits is 8.

What am I?

4400

Money problems 1

1 Stacey's mum writes the correct amount in words on a cheque: 'Seven thousand, eight hundred and two pounds'. In figures she writes 78002.

Explain why she is wrong.

Write the amount correctly. **£7802**

2 Tariq spends £56 on trainers and £97 on a coat.

How much does he spend altogether? **£153**

3 The price of a tube of Jitters is 29p. For a school party Mr Wilks buys a box of 100 tubes.

How much in £ and p does Mr Wilks pay? **£29**

4 Katie pays £36 for her two dogs and £85 for her 5 cats to enter a pet show. She wins £150 in prize money.

How much has she won when she deducts the cost of entering her pets? **£29**

5 It costs Paul's dad £44 to take them both to see a show. His dad's ticket cost £27.

How much did Paul's ticket cost? **£17**

6 A school raises £150 for six charities. Each charity is given one-sixth of the amount raised.

How much did each charity get? **£25**

7 I am half-way between double 17 and a half of 100.

What am I?

42

5

Number problems 2

❶ A museum has 19 very old dolls. Twelve of them are taken away to be repaired.

How many dolls are left in the museum? **7**

❷ A stall has 84 fancy bracelets for sale. By midday half of them have been sold. By 3 p.m. half of those remaining have been sold.

How many bracelets are left at 3 p.m? **21**

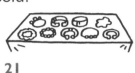

❸ For a school party Mrs Bright buys six boxes of straws.

How many straws does she get in the six boxes? **750**

❹ A school car park has places for 18 cars. The teachers use 11 of the places. During the day five more cars are parked.

How many empty places are left? **2**

❺ Each jar has 240 vitamin tablets.

How many vitamins are in a box of 8 jars? **1920**

❻ A waiter puts out 63 cups and saucers one morning ready for the day.

How many pieces of crockery does he put out? **126**

❼

I am a multiple of both 3 and 7. The difference between my two digits is 3.

What am I?

63

6

Number problems 3

1 Seven apple pies are cut into eight equal pieces. Of these pieces, 37 are eaten.

Write the amount of apple pies left as a mixed number. **$2\frac{3}{8}$**

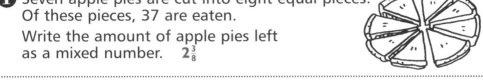

2 A trailer when full carries 9 motor bikes. A van with a full trailer makes three journeys to deliver motor bikes to a shop.

How many motor bikes does the shop get? **27**

3 On a Snakes and Ladders board Lucas has got to number 76. To win he needs to get to 100.

Altogether how many more does he need to throw to win? **24**

4 A motel has 41 rooms on the ground floor and 59 rooms on the first floor.

How many rooms does the motel have altogether? **100**

5 A baker makes a very large cake. He cuts it into 10 equal strips. He puts 7 of the strips onto a plate.

What fraction of the cake is on the plate? **$\frac{7}{10}$**

He cuts the 7 strips into tenths.

How many hundredths of the cake are on the plate? **70**

6 In a packet there are 18 chocolate biscuits. The biscuits are shared equally on three plates.

How many biscuits are on each plate? **6**

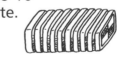
18 biscuits

7

I am two numbers. The sum of my two numbers is 100. The difference between my two numbers is 24.

What am I?

38, 62

Number problems 4

1 Newly born chicks are sent to a farm; 350 of them go in a van and 650 in a trailer.

Altogether how many chicks go to the farm?
1000

2 For every four tokens from packets of Rice Pops you can get one Demetrius monster.

How many monsters can you get with 40 tokens? **10**

3 Around each circular table there is room for 5 children.

How many children can be seated around 5 tables?
25

4 Each pack has 6 drink cartons.

How many cartons are there in four packs? **24**

5 At the start of a day a shop has 100 books of first class stamps. It sells 72 books.

What percentage of the books is sold? **72%**

What percentage of the books is left? **28%**

First class stamps

6 A free bottle of pop is given away with every 20 tokens from bags of crisps. Jennifer has 100 tokens.

How many free bottles of pop can she get? **5**

FREE POP

7  My first digit is odd. My second digit is even and less than my first digit. The product of my two digits is 18.

What am I?

92

8

Number problems 5

1 A bus has two seats down each side of an aisle. There are 9 rows of seats.
How many seats are there on the bus?   **36**

2 There are 98 yachts in a regatta. Because of high seas only half of the yachts complete the race.
How many yachts fail to complete the race?
49

3 A hockey team plays 10 games. It scores 0 goals in 1 game, 1 goal in 3 games, 2 goals in 4 games and 3 goals in 2 games.

How many goals did they score in the ten games?   **17**
What was the mode of their scores?   **2**

4 A bricklayer's hod can hold 8 bricks. He goes up a ladder five times with his hod full of bricks.

Altogether how many bricks does he carry?   **40**

5 At the afternoon performance of a school play there are 74 parents. The teachers hope that double that number will come in the evening.
How many parents do the teachers hope will come in the evening?   **148**

6 A box of 32 new paint brushes arrives at school. The brushes are divided equally between four classes.
How many paint brushes does each class get?   **8**

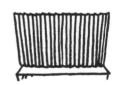

7

I am a three-digit number. I am an odd multiple of 50. The product of my first two digits is 45.

What am I?

950

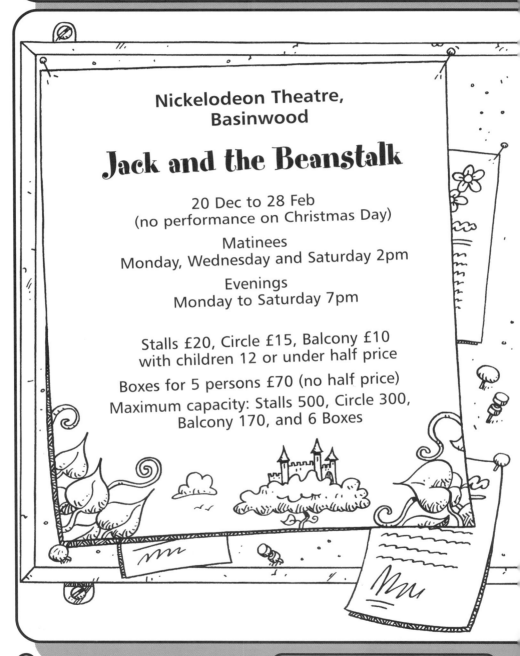

Nickelodeon Theatre, Basinwood

Jack and the Beanstalk

20 Dec to 28 Feb
(no performance on Christmas Day)

Matinees
Monday, Wednesday and Saturday 2pm

Evenings
Monday to Saturday 7pm

Stalls £20, Circle £15, Balcony £10
with children 12 or under half price

Boxes for 5 persons £70 (no half price)
Maximum capacity: Stalls 500, Circle 300,
Balcony 170, and 6 Boxes

1 On which day of the week is there no performance? **Sunday**

2 How many performances are there in a complete week? **9**

3 On how many days is there at least one performance? **6**

4 When all of the six boxes are full what is the total number in them? **30**

5 On Monday afternoon only half of the seats in the theatre are occupied. How many people were watching the pantomime? **500**

6 On 2nd January the theatre is full. 250 of those not in boxes have paid full price. How many children have paid half price? **720**

7 Each performance lasts for $2\frac{1}{2}$ hours, including an interval of 20 minutes. At what time will the evening performance end? **9:30 p.m.**

8 Five friends book a box. They share the cost equally between them. How much does each one of them have to pay? **£14**

9 On 12th February the balcony is full. There are 100 children under 12 in the balcony. What is the total amount that the people in the balcony have paid? **£1200**

10 Mr and Mrs Blue take their three children, all under 12, to see the pantomime. They sit in the stalls. How much do they pay for the whole family? **£70**

11 An ice cream costs £1.50 and a drink £1.10. Two adults have an ice cream each and four children a drink each. What is the total cost? **£7.40**

12 There are three different teams of ten children dancers. What is the total number of children in the three teams? **30**

13 The beanstalk is 3.5 metres in length. Jack climbs down from the top and jumps the last 2 metres. How far has he climbed down? **1.5 m**

Length problems 1

1 A jumbo jet flies 6782 miles from London to Singapore, and then 4317 miles to Sydney.

What is the total distance the jumbo flies?
11 099 miles

What is the distance rounded to the nearest 100 miles?
11 100 miles

2 A book is 15.5 cm wide and 23.5 cm long.

15.5 cm

23.5 cm

What are the measurements of the book in millimetres?
155 × 235 mm

How many centimetres is the perimeter of the book?
78 cm

3 A Victorian table in a museum measures 5 m by 90 cm.
What is the perimeter of the table in centimetres?
1180 cm

4 The length of a runway is 1.8 km. A plane starts at one end of the runway and takes off after 1.2 km.

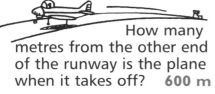

How many metres from the other end of the runway is the plane when it takes off? **600 m**

5 After an 800 m race the winner is 1.3 metres ahead of the second place runner.

By how many centimetres did she win? **130 cm**

6 Chumo builds a table top 80 cm square. He glues fancy ribbon around the four edges of the square top.
What is the length of the ribbon? **3.20 m**

7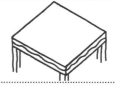

The product of my two digits is 16. My first digit is 4 times my second digit.

What am I?

82

12

Time problems 1

1 Paul leaves home at 09:18. He arrives at his destination at 13:53.

How long has his journey taken him?
4 hr 35 min

2 In a practice a Formula One car does seven laps around the track. Each lap takes 1 min 14 sec.

How long does it take the car to do the seven laps?
8 min 38 sec

3 Marvin has two minutes to catch the bus. He runs for 49 seconds and sees that the bus has not come so he walks for the rest of the way. It takes him 1 min 37 sec altogether.

For how many seconds did he walk? **48 sec**

4 A train leaves the station at 11:42. It stops after 2 hr 30 min.

What is the time when the train first stops? **2:12 p.m. or 14:12**

5 A rowing crew take 43 minutes to row the first half of a race. They row the second half in 38 minutes.

How long in hr and min did it take them to row the whole race? **1 hr 21 min**

6 A plane flies from London to Manchester in 53 minutes. It does the journey there and back four times a day.

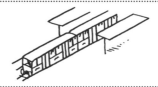

For how many hr and min is the plane in the air?
7 hr 4 min

7

I am a multiple of 2, 5 and 7. The sum of my three digits is 10. I am not 280.

What am I?

910

Measures problems 1

1 At a supermarket Neil weighs out 5 kg of apples. He then decides he wants another 7 kg.

How many kilograms has he weighed altogether?   **12 kg**

2 A large suitcase has a capacity of 111 litres.

What is the capacity of two of the suitcases?
222 L

3 Leon usually takes 75 minutes to travel by car to his friend's house. On Monday, because of road works, it takes him twice as long.

How long does it take him on Monday?   **2 hr 30 min**

4 A tree is 17 metres tall. After high winds it becomes dangerous and 8 metres are cut off its top.

What is the length of the tree after it is cut?   **9 m**

5 A bottle of 300 mL of ketchup weighs 342 g.

How many grams will a 600 mL bottle of ketchup weigh?   **684 g**

6 Sophie has a 5 m by 6 m rectangular lawn. She digs up half of the lawn to make a vegetable patch.

What is the area of the vegetable patch?   **15 m²**

7

I am a decimal number. Count on in 0.5s from zero and you will get to me immediately before you say 10.

What am I?

9.5

Money problems 2

1 Reubin buys both the computer and the printer.

What is the total cost?
£924

2 Dawn's mum has saved £370 for a second-hand car. She sees one she likes, but it is double what she has.

What is the price of the car Dawn's mum has seen?
£740

3 Jo pays £1.80 for three custard tarts.
What is the price of one custard tart? **60p**

4 In a sale a washing machine costs £399.50. Its price has been reduced by £130.
What was its price before the sale?
£529.50

£130 OFF
£399.50

5 A family of three adults and three children visit the circus.
What is the total cost? **£24**

Adults £5.50
Children £2.50

6 Tim buys the fridge freezer for £385 instead of £402.
How much money does he save? **£17**

£402

7 My digits are the first three numbers in the count on in 3s sequence.
My hundreds digit is three times my unit digit.

What am I?

963

15

Number problems 6

1 A trailer can carry 24 bales of hay. The farmer makes four journeys with full trailers.

Altogether how many bales does the farmer take? **96**

2 A cruise ship can take 1824 passengers; 5617 people apply for the cruise.

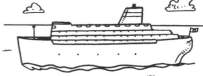

How many people will not get on the cruise? **3793**

3 A baker makes 120 cakes. He displays them in the window of his shop equally on three trays.

How many cakes are on each tray? **40**

4 A DIY superstore has a two week sale. In the first week it sells 3294 tins of paint. In the second week it sells 486 fewer than in the first week.

How many tins of paint does the store sell in the two weeks?
6102

£9.99

5 A multi-storey car park has five levels. Each level can hold 200 cars.

How many cars are in the car park when it is full?
1000

6 A joiner makes four-legged stools. First of all he makes 96 legs for the stools.

How many stools is the joiner intending to make? **24**

7

I am the tenth number that appears in both the count on in 4s and count on in 6s sequences.

What am I?

120

Number problems 7

1 In a city centre the temperature at 19:00 is 18°C. Every hour until 02:00 the next day the temperature drops 3°C.
What is the temperature at 02:00?   **−3°C**

2 A box holds 24 cartons of juice. Ten boxes are stacked on a supermarket's shelves.
How many cartons are on the shelves?
240

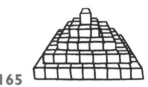

3 Clive uses cubes to make a step pyramid. He makes a 9 by 9 cube square base. On the next level he puts 7 by 7 cubes, then 5 by 5 cubes and so on.
How many cubes does he use altogether?   **165**

4 Jenny buys ten full books of first class stamps. She works out that there are 120 stamps altogether.
How many stamps are in each book?   **12**

5 A transport company has 40 lorries. Each lorry has six tyres. All the tyres on every lorry have to be replaced as they are faulty.
How many tyres are replaced?   **240**

6 In one week 84 stray cats are brought to a rescue centre. The cats are put six to a cage.
How many cages are used?   **14**

7 I am an odd multiple of 25. My three digits are all odd. The sum of my three digits is 15.   What am I?

375

17

Review problems 2

1 In 1800 a royal barge had 21 oars on each side, with 4 oarsmen to each oar.

How many oarsmen were there altogether? **168**

2 Carmel buys new jeans. The legs of the jeans are 83 cm long. She shortens the legs by 35 mm.

What are the lengths of the legs when shortened? **79.5 cm**

3 Zoe buys 3.5 kg of Granny Smiths and 2.5 kg of Red Delicious apples.

What is the total weight of apples that Zoe buys? **6 kg**

4 Keiron programmes his video to start at 09:50 and to end at 12:15.

For how long does the video record? **2 hr 25 min**

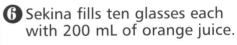

5 Glyn gets £3.50 pocket money each week. He is saving to buy a £40 model car.

For how many weeks will he need to save to buy the car? **12**

6 Sekina fills ten glasses each with 200 mL of orange juice.

How many litres of orange juice does she use? **2 L**

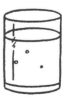

7  I am an even multiple of 20. I am less than 300. The sum of my digits is 7. What am I?

160

Number problems 8

1 In a fishing port there are 63 in-shore fishing boats and 37 sea-going trawlers.

What is the total number of fishing boats in the port?   **100**

2 In a park there are 9 swings. The playground is enlarged to make room for twice as many swings.

How many swings are in the playground now?   **18**

3 There are 19 tea bags left in the jar. Colin makes tea with three of them.

How many tea bags are left after Colin has made tea?   **16**

4 At 20:00 Jessica sees that the thermometer reads 5.5°C. At midnight it reads −1.5°C.

5.5°C

How many degrees Celsius has the temperature dropped since 20:00?   **7°C**

5 Les shares a packet of 28 chocolate drops fairly with his sister.

How many chocolate drops does he give his sister?   **14**

6 A cafe has eleven people seated at tables. Eight more people come into the cafe and sit down.

How many people are sitting in the cafe then?   **19**

7 Divide me by 4 and multiply the answer by 6 and you get 72. What am I?   **48**

Money problems 3

1 Michelle pays £2.50 for a magazine and £7.50 for a book.

What is the total amount she pays? **£10**

2 Kashif's father buys a new car. He writes a cheque for £19 502.

How does he write this amount in words?

Nineteen thousand, five hundred and two pounds

3 Euan buys two packets of QU tea bags.

£3.98

Buy 2 for £7

How much does he save on the total cost of two separate packets? **96p**

4 Apples cost £1.09 per kg. Elaine buys 3 kg.

What is the total cost of the 3 kg of apples? **£3.27**

Apples

£1.09 per kg

5 Rick buys both the Cheese Wavers and the French Fries crisps.

How much does the special offer save him? **52p**

Buy both for £2.75

£1.58 CHEESE WAVERS

£1.69 FRENCH FRIES

6 Vince buys the pack of 3 cans. Roger does not see the offer and buys three separate cans.

How much less does Vince pay than Roger? **15p**

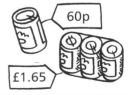

60p

£1.65

7 I am a three-digit multiple of 10. I am more than 600. The sum of my first two digits is 13. The difference between my first two digits is 3.

What am I?

850

Number problems 9

❶ There are 53 stalls selling goods at an open-air market. After a violent storm 25 stalls close.

How many stalls remain open?   **28**

---

❷ Each box has 12 peaches.

How many peaches are in a case of 25 boxes?   **300**

---

❸ In a dog van there is room for five dogs each in its own cage. In all 35 dogs have to be moved to new kennels.

How many trips will the van have to make to take all the dogs?   **7**

❹ It is Craig's job to inspect teddy bears in a factory in case they are faulty. He checks 19 teddy bears every 10 minutes.

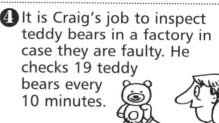

How many does Craig inspect in an hour?   **114**

---

❺ At a restaurant 64 places are set at tables each with a knife, fork and spoon. There are 8 tables.

Altogether how many pieces of cutlery do the waiters have to put out?   **192**

❻ Geri is taking part in a golf competition. The scores in her two rounds are 87 and 79.

What is Geri's total score for the two rounds?   **166**

---

❼

I am two numbers. The sum of my numbers is 100. The product of my numbers is 1275.

What am I?

15, 85

Number problems 10

1 In a packet of 60 party balloons one-tenth are red.

How many balloons are red? **6**

What fraction of the balloons are not red? $\frac{9}{10}$

60
balloons

2 Every day from Monday to Friday Natasha bakes 20 tarts.

How many tarts does she bake in the five days? **100**

---

3 In a packet there are 120 tea bags. The packets are delivered to shops in boxes of 100 packets.

How many tea bags are there in a box? **12 000**

120 TEA BAGS
100 PACKETS

---

4 A sale has 700 polo shirts. On the first day one-hundredth of the shirts are sold.

How many shirts are sold on the first day? **7**

5 A car park has ten levels. On each level it is possible for 217 cars to park.

How many cars are in the car park when it is full? **2170**

FULL

---

6 A survey finds that the average family has 1.7 children.

What is 1.7 as a mixed number? $1\frac{7}{10}$

EVENING POST
SURVEY of FAMILY SIZE

---

7 I am divisible by 4 and 10. The difference between my two digits is 8.

What am I?

80

Measures problems 2

degrees; halves; decimals total 10; ×/÷6

❶ Marcus works in a restaurant. He weighs 83 kg of flour. He uses half of it to make bread.

What is the weight of the flour in kg and g that he used for the bread?   **41 kg 500 g**

---

❷ A plane is travelling north-east. The pilot changes direction so as to travel north-west.

Through how many degrees does the plane turn?   **90°**

❸ A 6-pack of cans of blackcurrant juice altogether contains 1.5 litres.

1.5 litres

How much juice is in each can?   **250 ml**

---

❹ The hands of a clock move from 1 o'clock to 2 o'clock.

How many degrees has the small hand turned through?   **30°**

How many degrees has the large hand turned through?   **360°**

---

❺ A photograph is 15 cm wide and 10 cm high.

← 15 cm →

10 cm

What is the area of the photograph?   **150 cm²**

❻ A telegraph pole is 10 m long. A length of 5.3 m is cut off.

What is the length of the pole that is left? **4.7 m**

---

❼

I am a two-digit number. When I am divided by 2, 3, 4 or 5 there is always a remainder of 1.

What am I?

61

Review problems 3

Comet
length 35.97 m
wing span 34.98 m
weight (loaded)
73 500 kg
ceiling 11 900 m
range 6900 km

Vulcan
length 30.45 m
wing span 33.83 m
weight (loaded)
77 300 kg
ceiling 19 810 m
range 7400 km

Canberra
length 19.95 m
wing span 19.5m
weight (loaded)
20 900 kg
ceiling 14 600 m
range 4275 km

Adults £5.50

Senior Citizens £3.50

Children under 16 £3.50

Children under 5 Free

Teachers' Notes

Content

Introduction

The Pupils' books

Word Problems, a series of four books, one for each of the years 3 to 6, helps develop children's ability to solve number problems in a variety of contexts. Each page of word problems may be used to support numeracy lessons taught earlier in a week or for homework. The books provide weekly practice of word problems, as described in the National Numeracy Strategy (NNS) *Framework* and are designed to match the weekly structure within the NNS "sample medium-term plans". The style of questions reflects the examples that appear in the National Curriculum tests.

The books contain two types of page:

Topic pages: Each page has six word problems which are devoted solely to one Topic (Number, Money, Time, Length, Weight, Capacity and Measures). The overall mathematical content of all the Topic pages is listed on page iii. The mathematical content of the word problems on each page is listed in the lozenge at the top of the page.

Question 7 on each Topic page is a number puzzle, which gives children practice in Reasoning with Number using a variety of number properties. These questions are similar to those in the *Framework* and the National Curriculum tests.

Review pages look back at the mathematical content of previous Topic pages. Some Review pages are double page spreads, with the left page having a scene with information that is required to answer questions, and the right page asking questions about the scene.

Illustrations are used to tune children in to the 'real life' context of the word problem. Each is part of the problem and it is important that children look at the illustrations closely, they may contain information that is also in the word problem itself. On some occasions an illustration will contain information that is not in the text of the word problem, but is essential in order to solve the problem.

Answers to word problems are printed in red at the side of each question in the pupils' pages of the Teachers' notes.

Helping children solve word problems

Here are some suggestions for helping children develop a strategy for solving word problems.

- Make sure children read a question carefully and not merely search for key words such as 'altogether' which they think, sometimes incorrectly, tell them what to do with the numbers.

- Encourage children to close their eyes and picture the context of a problem and any actions that are performed with/on 'objects' in the context.

- Allow children to talk with other children about a problem and to ask themselves:
  - 'What do I have to find out?'
  - 'What do I know that will help me find out?'
  - 'What do I have to do with what I know to find out?'

It is also important as part of a word problem solving strategy that children develop:

- the skill of recognising information that is helpful and that which is not helpful when trying to solve a word problem.

- the ability to choose and use appropriate operations to solve word problems.

These can only develop with experience of solving many word problems over a long period of time.

Calculators

In Year 5 of the NNS Framework children are expected to decide when it is appropriate to calculate mentally, with written support or to use a calculator. You will need to discuss with your children which of the methods they feel are suitable for different word problems and have calculators available when they feel the need to use them. However, it is important that you encourage them to try to solve word problems either mentally or with written support as much as possible. This will differ from child to child.

Teaching plans

So that you can integrate *Word Problems* into your medium-term teaching plans pages 4 to 6 show the relationship between the Topics in *Word Problems* and the Sample Medium-term plans suggested by the NNS. Pages 7 and 8 show how the Topics in *Word Problems* relate to Mathematics in the National Curriculum in Wales Programme of Study and the similar Programme of Study for Northern Ireland.

Summary of mathematical content

page	topic	Mathematical content
4	Number 1	compare; round; ×/÷2/10/100
5	Money 1	write numbers; +/−2D numbers; ×100; 1/6 of
6	Number 2	+/− facts to 20; doubles/halves; HTU×U
7	Number 3	fractions; mixed numbers; ×/÷3; pairs total 100
8	Number 4	ratio; percentages; ×4 & 5; multiples of 50 total 1000
9	Number 5	doubles/halves; ×/÷4 & 5; mode
10–11	**Review 1**	**review of previous content**
12	Length 1	miles; convert units; round; perimeter; ThHTU+ThHTU
13	Time 1	24 hr clock; +/−2D numbers; ×7 & 8
14	Measures 1	+/− facts to 20; doubles/havles; ×/÷2
15	Money 2	small differences; doubles; ×/÷3; HTU+HTU
16	Number 6	ThHTU+/−ThHTU; ×/÷4 & 5
17	Number 7	-ve numbers; square numbers; ×/÷6 & 10
18	**Review 2**	**review of previous content**
19	Number 8	-ve numbers; +/− facts to 20; pairs total 100; ×/÷2
20	Money 3	write numbers; ×/÷2 & 3; multiples of 50 total 1000
21	Number 9	+/−2D numbers; ×12 & 19; ÷5 & 8; HTU÷U
22	Number 10	fractions/decimals; ×5, 10 & 100
23	Measures 2	degrees; halves; decimals total 10; ×/÷6
24–25	**Review 3**	**review of previous content**
26	Measures 3	+/− facts to 20; area; ThHTU-ThHTU
27	Weight 1	four operations
28	Number 11	near doubles; ThHTU+ThHTU; ×/÷2 & 4
29	Money 4	+/− near multiples of 10/100; ×/÷3
30	Number 12	×/÷7, 8 & 9
31	Number 13	multi-step with four operations
32–33	**Review 4**	**review of previous content**
34	Number 14	-ve numbers; +/− facts to 20; pairs total 100; round
35	Money 5	multiples of 50 total 1000; ÷2, 5 & 10, round up/down
36	Number 15	facts/pv to ×/÷; +/−2D numbers; TU×TU
37	Number 16	fractions/decimals; percentages; ×/÷10
38	Number 17	ratio; proportion; percentages; ×/÷7
39	Number 18	mode; range; ×/÷8 & 9
40–41	**Review 5**	**review of previous content**
42	Time 2	24 clock; percentages; ×/÷6
43	Capacity 1	pints/gals; decimals total 1; +/−2D numbers; ×/÷7
44	Measures 3	decimals total 10; facts/pv to −; ×/÷8 & 100
45	Money 6	+ several numbers; pairs total 100; ×2 & 3; +/−£s & p
46	Number 19	facts/pv to +/−; +/−decimals; ThHTU+/−ThHTU
47	Number 20	multi-step with four operations
48	**Review 6**	**review of previous content**

Word Problems and the National Numeracy Strategy
Sample medium–term plans

AUTUMN			
Sample medium-term plans		**Word Problems**	
Unit	**Topic**	**Pages**	**Topic**
1	Place value; ordering; rounding Using a calculator	4	Number 1
2–3	Understanding × and ÷ Mental calculation strategies (× and ÷) Paper and pencil procedures (× and ÷) Money and 'real life' problems Making decisions; checking results including using a calculator	5–6	Money 1 Number 2
4–6	Fractions, decimals and percentages Ratio and proportion	7–8	Number 3 Number 4
6	Handling data Using a calculator	9	Number 5
7	**Assess and review**	**10–11**	**Review 1**
8–10	Shape and space Reasoning about shapes Measures, including problems	12–14	Length 1 Time 1 Measures 1
11	Mental calculation strategies (+ and −) Pencil and paper procedures (+ and −) Money and 'real life' problems Making decisions; checking results, including using a calculator	15–16	Money 2 Number 6
12	Properties of numbers Reasoning about numbers	17	Number 7
13	**Assess and review**	**18**	**Review 2**

SPRING			
Sample medium-term plans		*Word Problems*	
Unit	Topic	Pages	Topic
1	Place value; ordering; rounding Using a calculator	19	Number 8
2–3	Understanding × and ÷ Mental calculation strategies (× and ÷) Paper and pencil procedures (× and ÷) Money and 'real life' problems Making decisions; checking results including using a calculator	20 21	Money 3 Number 9
4	Fractions, decimals and percentages Using a calculator	22	Number 10
5	Shape and space Reasoning about shapes	23	Measures 2
6	**Assess and review**	**24 and 25**	**Review 3**
7–8	Measures including problems Handling data	26 27	Measures 3 Weight 1
9–10	Mental calculation strategies (+ and ÷) Pencil and paper procedures (+ and ÷) Money and 'real life' problems Making decisions, checking results, including using a calculator	28 29 30	Number 11 Money 4 Number 12
11	Properties of numbers Reasoning about numbers	31	Number 13
12	**Assess and review**	**32 and 33**	**Review 4**

SUMMER			
Sample medium–term plans		Word Problems	
Unit	Topic	Pages	Topic
1	Place value, ordering, rounding Using a calculator	34	Number 14
2–3	Understanding × and ÷ Mental calculation strategies (× and ÷) Paper and pencil procedures (× and ÷) Money and 'real life' problems Making decisions; checking results including using a calculator	35–36	Money 5 Number 15
4–5	Fractions, decimals and percentages Ratio and proportion	37–38	Number 16 Number 17
6	Handling data Using a calculator	39	Number 18
7	Assess and review	40–41	Review 5
8–10	Shape and space Reasoning about shapes Measures, including problems	42–44	Time 2 Capacity 1 Measures 4
11–12	Mental calculation strategies (+ and −) Pencil and paper procedures (+ and −) Money and 'real life' problems Making decisions, checking results, including using a calculator	45–46	Money 6 Number 19
13	Properties of numbers Reasoning about numbers	47	Number 20
	Assess and review	48	Review 6

Word Problems and the National Curriculum in Wales

Using and Applying Mathematics
U1. Making and Monitoring Decisions to Solve Problems
U2. Developing Mathematical Language and Communication
U3. Developing Mathematical Reasoning
Number
N1. Understanding Number and Place Value
N2. Understanding Number Relationships and Methods of Calculation
N3. Solving Numerical Problems
Shape; Space and Measures
S2. Understanding and Using Properties of Position and Movement
S3. Understanding and Using Measures
Handling data
D1. Collating, Representing and Interpreting Data

Word Problems		Relevant Sections of the National Curriculum Programme of Study								
page	topic	U1	U2	U3	N1	N2	N3	S2	S3	D1
4	Number 1	x	x	x	x	x	x			
5	Money 1	x	x	x	x		x			
6	Number 2	x	x	x		x	x			
7	Number 3	x	x	x	x	x	x			
8	Number 4	x	x	x	x	x	x			
9	Number 5	x	x	x		x	x			x
10–11	Review 1	x	x	x		x	x		x	
12	Length 1	x	x	x	x		x		x	
13	Time 1	x	x	x			x		x	
14	Measures 1	x	x	x			x		x	
15	Money 2	x	x	x			x			
16	Number 6	x	x	x		x	x			
17	Number 7	x	x	x	x	x	x			
18	Review 2	x	x	x		x	x		x	
19	Number 8	x	x	x	x	x	x			
20	Money 3	x	x	x	x		x			
21	Number 9	x	x	x		x	x			
22	Number 10	x	x	x	x	x	x			
23	Measures 2	x	x	x	x		x	x	x	
24–25	Review 3	x	x	x		x	x		x	
26	Measures 3	x	x	x			x		x	
27	Weight 1	x	x	x			x		x	
28	Number 11	x	x	x		x	x			
29	Money 4	x	x	x			x			
30	Number 12	x	x	x		x	x			
31	Number 13	x	x	x		x	x			
32–33	Review 4	x	x	x		x	x		x	
34	Number 14	x	x	x	x	x	x			
35	Money 5	x	x	x			x			
36	Number 15	x	x	x		x	x			
37	Number 16	x	x	x	x		x			
38	Number 17	x	x	x	x	x	x			
39	Number 18	x	x	x			x			x
40–41	Review 5	x	x	x		x	x		x	
42	Time 2	x	x	x	x		x		x	
43	Capacity 1	x	x	x			x		x	
44	Measures 4	x	x	x			x		x	
45	Money 6	x	x	x			x			
46	Number 19	x	x	x		x	x			
47	Number 20	x	x	x		x	x			
48	Review 6	x	x	x		x	x		x	

Matching *Word Problems* to The National Curriculum in Northern Ireland

PROCESSES IN MATHEMATICS
P1. Using Mathematics
P2. Communicating Mathematically
P3. Mathematical Reasoning
SHAPE AND SPACE
S2. Position, Movement and Direction
HANDLING DATA
D1. Collect, Represent and Interpret Data

NUMBER
N1. Understanding Number and Number Notation
N2. Patterns; Relationships; and Sequences
N3. Operations and their Applications
N4. Money
MEASURES (M)

Word Problems		*Relevant Sections of the National Curriculum Programme of Study*									
page	topic	P1	P2	P3	N1	N2	N3	N4	M	S2	D1
4	Number 1	x	x	x	x		x				
5	Money 1	x	x	x	x			x			
6	Number 2	x	x	x			x				
7	Number 3	x	x	x	x	x	x				
8	Number 4	x	x	x	x		x				
9	Number 5	x	x	x			x				x
10 and 11	**Review 1**	x	x	x		x	x	x	x		
12	Length 1	x	x	x	x				x		
13	Time 1	x	x	x	x				x		
14	Measures 1	x	x	x		x			x		
15	Money 2	x	x	x	x	x		x			
16	Number 6	x	x	x			x				
17	Number 7	x	x	x		x	x				
18	**Review 2**	x	x	x			x	x	x		
19	Number 8	x	x	x			x				
20	Money 3	x	x	x	x			x			
21	Number 9	x	x	x			x				
22	Number 10	x	x	x	x		x				
23	Measures 2	x	x	x	x	x			x	x	
24 and 25	**Review 3**	x	x	x	x		x	x	x		
26	Measures 3	x	x	x					x		
27	Weight 1	x	x	x					x		
28	Number 11	x	x	x		x	x				
29	Money 4	x	x	x			x	x			
30	Number 12	x	x	x			x				
31	Number 13	x	x	x			x				
32 and 33	**Review 4**	x	x	x			x	x	x		
34	Number 14	x	x	x	x						
35	Money 5	x	x	x		x	x	x			
36	Number 15	x	x	x			x				
37	Number 16	x	x	x	x		x				
38	Number 17	x	x	x	x		x				
39	Number 18	x	x	x			x				x
40 and 41	**Review 5**	x	x	x				x	x		
42	Time 2	x	x	x	x				x		
43	Capacity 1	x	x	x					x		
44	Measures 4	x	x	x		x			x		
45	Money 6	x	x	x				x			
46	Number 19	x	x	x			x				
47	Number 20	x	x	x			x				
48	**Review 6**	x	x	x			x	x	x		

1 Which aircraft has the greatest length?   **Comet**
How many metres longer is it than the Canberra?   **16.02 m**

2 Approximately how many times heavier is the Vulcan than the Canberra when both are fully loaded?   **3 or 4 times**

3 The weight of the Comet when empty is 36 100 kg. What is the greatest weight of passengers and cargo it can carry?   **37 400 kg**

4 How much higher can the Canberra fly than the Comet?   **2700 m**

5 The Vulcan can double its range by refuelling in the air. How far could it then fly?   **14 800 km**

6 The take-off distance for the Comet is 2270 metres. Could the Comet use a runway 1.5 miles long? Explain how you decided.

7 Write the wing spans of the 3 aircraft in order, shortest first.
19.5 m, 33.83 m, 34.98 m

8 What is the difference, in centimetres, in the wing spans of the Comet and the Vulcan?   **115 cm**

9 The maximum speed of the Canberra was 930 km per hr. The maximum speed of the Vulcan was 100 km per hr more than this. What was the maximum speed of the Vulcan?   **1030 km/hr**

10 The Vulcan first flew on 30th August 1952. The Comet first flew on 27th July 1949. How many years and days was it that the Vulcan flew after the Comet did? **3 years 34 days**

11 Mr and Mrs Rich with their two children, Carrie aged 4 and Simon aged 10, and their Gran, a senior citizen, visit the museum. How much do they pay altogether?   **£18**

12 The Rich family left home at 09:15 and arrived at the museum at 11:05. How long did it take them to get to the museum?   **1 hr 50 min**

Measures problems 3

1 Danny and Emma together pick 11 kg of strawberries. Their mum and dad only pick 8 kg. Altogether how much does the family pick?
19 kg

2 A glass window is 0.6 m wide and 1 m high.

What is the area of the glass?
0.6 m²

1 m

← 0.6 m →

3 Sinead runs in a 5000 m race. After 3250 m she has to stop as she has an injury.

How far had she still to run?   **1750 m**

4 Gareth swims 400 m in 7 min 8 sec and then another 400 m in 8 min 9 sec.

How long did it take him to swim 800 m?   **15 min 17 sec**

5 A floor is 6 m long and 4 m wide. A carpet is put on the floor so that it is 1 metre from the edge of the room all around.
What is the area of the carpet?
8 m²

4 m

6 m

6 Carrie buys a carton of the apple juice. She drinks 300 mL of the juice.

How much of the juice is left?
700 mL

1 litre

7    I am half-way between double 150 and half of 800   What am I?

350

Weight problems 1

1 A jar of jam weighs 454 g. Francis uses 60 g of the jam to make tarts.

What is the weight of the jam left in the jar? **394 g**

2 A bakery makes 18 kg of biscuit dough and 17 kg of chocolate. They use them to make 1000 chocolate biscuits.

What is the weight in grams of each chocolate biscuit before it's baked? **35 g**

---

3 A bag has 6 packets of crisps. Each packet weighs 25 g.

What is the weight of the six packets?
150 g

6×25 g

---

4 A machine weighs 32 kg of Puffy Wheat. It then fills 100 boxes each with 160 g.

How much Puffy Wheat is left when all 100 boxes are filled? **16 kg**

160 g

---

5 There are 20 biscuits in the packet. What is the weight of one biscuit?
10 g

200 g

---

6 On a box of Pop-Pops it says, 'This package contains 15×30 g servings.' What is the total weight of the Pop-Pops in the box? **450 g**

This packet contains
15×30 g
servings

7

I am divisible by 8 and by 9.
I am a three-digit multiple of 10.
I am between 200 and 700.

What am I?

360

Number problems 11

❶ Every parent who comes to the school fair brings an adult friend with them. In all, 190 parents come to the fair.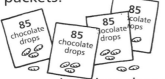

What is the total of parents and friends who come to the fair? **380**

❷ On a Tuesday 3851 passengers leave the airport on planes, and 2469 passengers arrive on planes.

How many passengers pass through the airport on the Tuesday? **6320**

❸ In every packet there are 85 chocolate drops. Sam buys four packets.

How many chocolate drops does Sam buy? **340**

❹ After two rounds of golf Sean's total was 152. He had the same score for each round.

What was Sean's score for each round? **76**

❺ At a wholesale market a shopkeeper buys two boxes of pears. One box has 350 pears, the other has 360.

How many pears does the shopkeeper buy altogether? **710**

❻ A garage replaces all the four tyres on a number of cars. Altogether it uses 92 of its stock of tyres.

How many cars have their tyres replaced? **23**

❼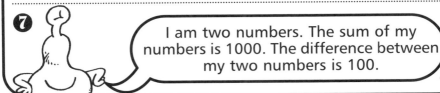

I am two numbers. The sum of my numbers is 1000. The difference between my two numbers is 100.

What am I?

450, 550

Money problems 4

1 Nyla buys a half a kilogram of grapes and a 6-pack of blackcurrant juice.
What is the total cost? **£1.78**

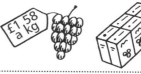

£1.58 a kg

99p

2 A mathematics instrument set costs £1.99. Mrs Williams buys one each for Ros and her two brothers.
How much does it cost Mrs Williams? **£5.97**

£1.99

3 Dale buys a calculator. He pays with a £10 note.
How much change does Dale get? **£3.01**

£6.99

4 Dino's dad buys him a blazer and a pair of trousers for school.
How much change does his dad get from a £50 note? **£12.02**

£21.99

£15.99

5 Rod, Stan and Harry each buy a pair of shorts.
How much would they each have saved had they bought a 3-pack and shared the shorts? **32p**

£2.40

3-pack £6.24

6 Jodie buys a padded jacket for £23.99 and a skirt for £7.99.
What is the total cost? **£31.98**

£7.99

£23.99

7  I am a four-digit number with the digits in order. My first two digits are the first two odd numbers. My last two digits are the last two single-digit even numbers.

What am I?

1368

29

Number problems 12

×/÷ 7, 8 and 9

1 There are seven letters and numbers on a car number plate.

AJ51XYZ

How many letters and numbers are there on six cars? **42**

2 A shop puts eight bunches of bananas on display. Each bunch has the same number of bananas. Altogether there are 56 bananas.

How many bunches of bananas are on display? **7**

3 Forty-two children line up in rows. There are seven children in every row.

How many rows do the 42 children make? **6**

4 A tractor is pulling a trailer. The trailer has eight bales on each level. There are four levels.

How many bales is the tractor pulling? **32**

5 A lift can carry a maximum of five people. A group are waiting to go in the lift. It goes up full nine times.

How many people are in the group? **45**

6 For a meeting 63 chairs are put into rows with nine chairs in each row.

How many rows of chairs are there? **7**

7 Divide me by 8 and multiply the answer by 7 and you get 35.

What am I?

40

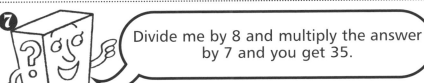

30

Number problems 13

1 At the start of its voyage a ferry has 1380 passengers. At its first stop 572 disembark and 317 get on.

How many passengers are on the ferry when it leaves? **1125**

2 A container has 34 tea bags. Two packets each with 95 tea bags are opened and put into the container.

How many tea bags are in the container after they are put in? **224**

3 In a pack there are 10 Jit-jats. For a party Kelly buys eight packs, and 63 Jit-jats are eaten.

How many Jit-jats are left at the end of the party? **17**

4 A shop sells packets of plane cards. There are 10 cards in a packet. Mrs Harrison buys nine packets, and divides the cards equally between her 3 children.

How many cards does each child get? **30**

5 Mr Best's garden has four planting areas. He already has 29 plants in his garden, but decides to buy another 43. He plants them so that there are the same number in each area.

How many plants are there in each area when he has finished planting? **18**

6 An airport has eleven piers for planes. Each pier has 10 planes. Half of the planes take off and 33 land.

What is the total number of planes at the piers after the ones that have landed arrive? **88**

7

Add 57 to me and you will get double 83.

What am I?

109

Travel Shop

Holiday in Spain
7 nights in Malaga at the Hotel Feria on the Costa del Sol for £309.
Children under 10 half price.

Holiday in **Portugal's** sunny Algarve at the Hotel Pacific for 7 nights £499.
Children under 12 half price.

Flights only
Alicante £79
Barcelona £139
Fuerteventura £139
Gibraltar £159
Ibiza £89
Lanzarote £109
Las Palmas £99
Malaga £79
Tenerife £79

Hotels only per night per room
Hotel Atlanta, Alicante £45
Hotel Oregon, Ibiza, £39
Hotel York, Lanzarote £37
Hotel Villa del Sol, Malaga £59
Hotel Santa Linacre, Tenerife £49

1 Mr and Mrs Wright have two children: Danny aged 13 and Teresa aged 8. Which holiday would be cheaper for them, the holiday to Malaga or the one to the Algarve?   **Malaga**

2 Frances's mum books for them both to stay at the Hotel Villa del Sol for 14 nights. They share a room. The extra 7 nights are half the price of the first 7 nights. How much do they pay?   **£619.50**

3 The two holidays in the window are too expensive for a retired couple. They look at a holiday in Alicante. For 7 nights in a 2-star hotel it is £168 per person and in a 3-star £250. How much less is it in the 2-star than the 3-star for the two of them?   **£164**

4 Ms Derry booked a flight to Ibiza. She also booked 10 nights at the Hotel Oregon. What did she pay? **£479**

5 Before her holiday Ms Derry changed £200 into Spanish pesetas. The exchange rate was 256.11 pesetas per £1. How many pesetas did she get?   **51 222 pesetas**

6 On her return she had 900 pesetas left. She exchanged them back at a rate of £1 for every 300 pesetas. How many £s did she get?   **£3**

7 The plane to Ibiza departed at 13:45. The flight time was 2 hr 50 min. At what time was the plane due to land in Ibiza?   **16:35**

8 When Ms Derry returned there were two Boeing 737s each taking 152 passengers. Altogether how many passengers did the two planes take?   **304**

9 The return flight for Ms Derry was at 23:10 on a Tuesday. The departure was delayed for 2 hours. At what time did the plane take off?   **01:10**

10 When Ms Derry arrived at the airport her luggage weighed 27 kg. This was 7 kilograms above the maximum allowed. What is the maximum weight of luggage?   **20 kg**

11 She was charged £10 for every extra kilogram. How much extra did she have to pay?   **£70**

Number problems 14

❶ The temperature at 5 a.m. outside Jamie's house is −4°C. By 12 o'clock the temperature has risen by 9°C.

9°C

What is the temperature at mid-day?   **5°C**

❷ Russell has six first class stamps. He buys a book that has 12 stamps.

How many stamps has he now?   **18**

❸ In a game of Snakes and Ladders Aisha is on square 73. She lands on a ladder that moves her up 14 places.

How many more places does she need to get to 100?   **13**

❹ At midnight the temperature at Abbey House is −3°C. At Castle Square it is 2°C.

How many degrees warmer is it at Castle Square than at Abbey House?   **5°C**

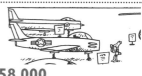

❺ Alice stands up 20 dominoes in a snake pattern. She pushes the end domino and 14 fall down.

How many dominoes are left standing?   **6**

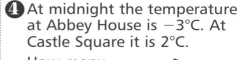

❻ In one year 275 639 people visit an aircraft museum. Of these 117 352 are children.

How many were adults?   **158 287**

Round this number to the nearest 1000.   **158 000**

❼

I am a multiple of 12. Round me to the nearest 10 and I become 70.

What am I?

72

Money problems 5

multiples of 50 total 1000; ÷2, 5 and 10, round up/down

1 For the family holiday Scott's mum buys the digital camera and the camcorder.

How much does she pay altogether? **£1000**

£350

£650

2 Tickets to a concert cost £2. Stacey has £21.

How many friends can Stacey take with her? **10**

♪ **£2.00**

3 Marie saves her money in ten different money boxes. She is given £37.50 for her birthday. She shares it equally between the ten boxes.

How much does she put into each box? **£3.75**

4 Adrian has saved £8.50 towards a front lamp for his mountain bike. The price of the lamp is £10.

£10.00

How much more does Adrian need? **£1.50**

5 The cost of using a mobile phone is 5p for every minute or part of a minute. Lee makes a 25 min 15 sec call.

How much does it cost him? **£1.30**

6 Ian has £37.26 in coins. He changes the coins into £5 notes.

How many £5 notes does he get? **7**

How much is left in coins? **£2.26**

7 I am an odd multiple of 50p. My number of £s is double 18.

What am I?

£36.50

Number problems 15

1 Sixty-nine hot air balloons take off. A further 87 are left on the ground. Later 38 of these take off.

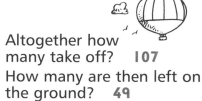

Altogether how many take off? **107**

How many are then left on the ground? **49**

2 A cinema has 600 seats. The film 'The XY People' has 40 screenings with all seats taken.

How many people see the film? **24 000**

3 An apple grower collects 5700 apples from her orchard. The apples are divided equally into 100 boxes to send to shops.

How many apples are in each box? **57**

4 A jet plane when full carries 63 passengers. It makes 36 flights and is full each flight.

What is the total number of passengers it carries on its 36 flights? **2268**

5 There are 650 adult guests at a charity dinner. A half of the guests are men.

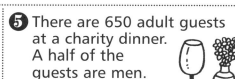

How many guests are women? **325**

6 There are 3 infant classes in a school. In each class there are 29 children.

How many children are in the infant school? **87**

7 Divide me by 5 and then divide the answer by 10. You will get 11.

What am I?

550

Number problems 16

1 A farmer has 300 sheep to load into a wagon. Before he stops for lunch he has loaded 30% of them.

How many does the farmer load before lunch? **90**

2 A group of loggers are set a target of cutting down 270 trees every week for 10 weeks.

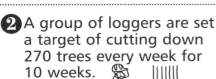

What is the total of trees they are expected to cut down in the 10 weeks? **2700**

3 One-half of the 200 oranges in a box are bad.

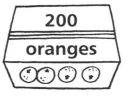

200

oranges

What percentage of the 200 oranges is bad? **50%**

4 A garden centre has 2500 daffodil bulbs for sale. An assistant puts them equally into 10 boxes.

How many daffodil bulbs are in each box?   **250**

🌸 SALE 🌸
Daffodil
bulbs
£5 for 25

5 In Year 5 there are 60 children. Three-quarters of them bring a packed lunch.

How many of Year 5 bring a packed lunch? **45**

6 One-quarter of the paints in a paint box are missing.

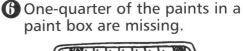

What fraction of the paints is still in the box? $\frac{3}{4}$

7

I am a number of tenths. My numerator is 3 less than my denominator.

What am I?

$\frac{7}{10}$

Number problems 17

1 Each pack has 10 ball point pens. Ennis buys seven packs to share with her friends at a party. How many ball point pens does she buy altogether?   **70**

2 Two in every three of the 300 people who visit a zoo are children. How many children visit the zoo? **200**

3 At a rounders cup final, 85% of the crowd are girls. What percentage are boys?   **15%**

4 At a tennis club there are 2 boys for every 3 girls. Altogether there are 6 boys in the tennis club. How many girls are in the tennis club?   **9**

5 At the Blackpool Tower 63 people are waiting to go up. The lift can only take 7 people. How many journeys does the lift make to carry all 63?   **9**

6 Tricia got 77 marks out of 110 in a maths test.

What percentage of the marks did Tricia get?   **70%**

7 I am a two-digit number. I am a multiple of both 7 and 9. What am I?

63

Number problems 18

❶ Eight hens each lay seven eggs.

Altogether how many eggs did they lay? **56**

❷ In ten throws of a ring Jake scores two 1s, one 2, two 3s and five 4s.

What is his total score? **30**

What is the mode of his scores? **4**

❸ A shop has nine fish tanks. The owner shares 81 fish equally between the nine tanks.

How many fish does he put into each tank? **9**

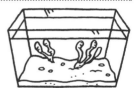

❹ The lowest score in a class test was 47. The highest score was 93.

What was the range of the scores? **46**

❺ The 64 children in Year 5 are divided into teams of eight.

How many teams are there? **8**

❻ There are nine teams in a relay race. Each team has four children.

How many children take part in the relay race? **36**

❼

I am a multiple of 7. Both my digits are even. The difference between my digits is 4.

What am I?

84

39

Review problems 5

Hanging baskets
Large   Small
£16   £12

Posts
£1.50 per metre

Usual price
£175
**Half price
in SALE**

Table & chairs
~~£480~~
20% off

Plants in pots
75p each

Paving stones
1 m × 1 m,
£3.95 each

Small plants
£2.99
box of 6

Edging stones
50 cm long
£1.50 each

Lawn mowers
Manual   Electric
£55.99   £79.99
Petrol
£199.99

Roses
£5.50
each

1 Mary's dad bought eight boxes of plants.
How much did he pay? **£23.92**
How many plants did he get? **48**

2 Kerry bought 3 pots of plants and 2 roses. She paid with a £20 note.
How much change did she get? **£6.75**

3 How many small hanging baskets can be bought for the price of three large ones? **4**

4 Jill's mum wants a lawn mower. How much more would she have to pay for an electric than a manual one? **£24**

5 What is the sale price of the garden shed? **£87.50**

6 Mr Sawyer bought the table and chairs. He paid with cash and was given an extra £50 off the sale price.
How much did he pay? **£334**

7 Sheila has a square lawn with sides 10 metres.
How many edging stones does she need to put around the lawn? **80**

8 Mr Ross has a rectangular area 5 m by 4 m that he wants to pave.
How many paving stones does he need? **20**
How much would they cost him? **£79**

9 Alice's dad wants 10 posts each 2 metres long.
How much would he pay for the posts? **£30**

10 A trolley can hold 10 boxes of plants.
Mrs Blakey buys 45 boxes.
How many journeys to her car will she have to make with the trolley? **5**

11 In the cafe muffins are 65p each.
How many can be bought with a £2 coin? **3**

41

Time problems 2

24 clock;
percentages; ×/÷6

1 Ayesha gets on to a bus at 11:53. She is meant to reach her destination at 12:49. Because of the traffic the bus arrives 5 minutes late.

How long does Ayesha's journey take?   **I hr I min**

2 A tram travels the length of Blackpool Promenade. It takes 37 minutes. The driver drives the tram six times along the promenade before resting.

For how long is the driver driving the tram?
3 hr 42 min

3 A train takes 6 hr 40 min to go from Plymouth to Newcastle. It leaves Plymouth at 09:33.

At what time is the train due to arrive in Newcastle?
16:13

4 Ricardo leaves for a holiday in Holland at 4.45 p.m. on Wednesday. He arrives at 10.45 a.m. the next day.

How long did the journey take him?   **18 hr**

5 It takes Beth five hours to get to her Gran's. She spends 80% of her time on a train.

How long does she spend on a train?   **4 hr**

6 Six bird watchers take turns to watch a nest for an equal amount of time. In total they watch for 10 hours.

How long does each person watch for?   **I hr 40 min**

7

Seventy-five per cent of me is 300.

What am I?

400

Capacity problems 1

pints/gals; decimals total 1; +/− 2D numbers; ×/÷7

1 A water butt contains 76 litres of water. After a long period of rain another 15 litres goes into the butt.
What is the total amount in the butt after the rain?   **91 L**

2 Marc makes 5 gallons of lemonade for a party. He puts the lemonade into pint glasses.
How many glasses can he fill with the 5 gallons?   **40**

1 pint

3 Gaby waters her plants. She uses a 0.5 litre jug. She fills it seven times with water.
What is the total amount of water she uses in L and mL?   **3 L 500 mL**

0.5 litres

4 There are 85 mL of vinegar left in a bottle. Becky uses 17 mL of it for cooking.
How many mL of vinegar is left in the bottle?   **68 mL**

284 mL

5 A bottle when full holds 1 litre. Simon puts 0.6 L of orange juice into the bottle.
How many more millilitres are needed to fill the bottle?   **400 mL**

1 litre

6 Joanne divides 0.7 L of apple juice equally between seven glasses.
How many millilitres of apple juice is in each glass?   **100 mL**

7
Both my numbers are multiples of 50. The sum of the numbers is 1000. One number is three times the other.

What am I?

250, 750

43

Measures problems 4

❶ A jar of beetroot weighs 340 g. With the vinegar removed it weighs 230 g.

net 340 g
drained
230 g

What was the weight of the vinegar in the jar? **110 g**

❷ Danny and Mel train for a marathon. They run 10 times around a 400-metre track.

How far do they run when training? **4 km**

❸ A case has eight bottles. The total capacity of the bottles is 24 litres.

How many litres does each bottle hold?
3 L

24 litres

❹ Jim buys a 3.8 kg bag of potatoes. His mother also buys 6.2 kg of potatoes.

Altogether what weight of potatoes have they bought? **10 kg**

3.8 kg   6.2 kg

❺ A school has eight lessons in a day. Every lesson lasts for 45 minutes.

What is the total number of hours that children spend in lessons? **6 hr**

❻ To make matchsticks a 5 metre length of very thin wood is cut into 100 matchsticks.

What is the length of each matchstick in centimetres?
5 cm

❼

The total of my two numbers is 99. Each digit of my two numbers is a 4 or a 5.

What am I?

45, 54

Money problems 6

+ several numbers; pairs total 100; ×2 and 3; +/− £s and p

1 Kerry buys both the clock and the watch. What is the total cost of the two items? **£40**

2 Rick buys a pack of steaks for £8.99 and a frozen duckling for £7.99. He uses the £1.50 off coupon.
What does he actually pay? **£15.48**

£1.50 OFF
when you spend over £15

3 Joe's mum is buying a car for £9936. She pays a deposit of £993.60.

How much has she left to pay when she collects the car? **£8942.40**

SUPER VALUE
£9936

4 Deidre buys two packets of fish fingers.

£1.99

What change does she get from a £5 note? **£1.02**

5 Hal buys the mini system and the turntable.
What is the total cost of the two? **£374.98**

£244.99
£129.99

6 Arthur collects Action Toys. He buys three new ones at £6.99 each.
What do the three cost him altogether? **£20.97**

£6.99 £6.99 £6.99

7

I am a multiple of 100. The sum of my four digits is 6. The product of my first two digits is 9.

What am I?

3300

Number problems 19

facts/place value to +/−;
+/− decimals;
ThHTU+/−ThHTU

❶ A jeweller has 473 watches for sale. A box arrives at the shop containing 200 watches.

200 watches

What is the total number of watches the jeweller has for sale after the box arrives?
673

❷ At an open air concert there are 8074 people. Of these 1368 were children.

How many of the spectators were adults? **6706**

❸ There are two judges at an ice skating competition.

4.7 **4.5**

Lucia scores 4.7 and 4.5. What is Lucia's total score?
9.2

❹ At a cricket final 5247 people support Aldwick and 2895 support Bertram. There are 562 who support neither team.

How many people are at the cricket match? **8704**

❺ There are 685 passengers waiting to board two Airbus jets to New York. In all, 340 of them get on one of the planes.
How many are left for the other plane? **345**

❻ Patrick is taking part in a diving competition. After his two dives he has scored 7.6 points. Darren has had one dive and scored 3.9 points.
How many does Darren need to score in his next dive to equal Patrick's total? **3.7**
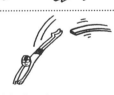

❼  I have four digits. Successive digits increase by 2. My tens digit is 6.

What am I?

2468

46

© REPP. Photocopying this material is not permitted.

Number problems 20

1 A garden centre has 153 roses for sale. Another 70 roses are delivered to the centre and put out for sale. Later 84 are sold.

How many roses are left after the 84 are sold? **139**

2 A farmer harvests his potatoes. He fills 100 bags each with 70 potatoes. He has 47 potatoes left.

How many potatoes did the farmer harvest? **7047**

3 Eight new wards are built at a children's hospital. The new wards share 72 beds equally between them. However, 4 beds in Ward 17 are sent back as they are faulty.

How many beds are left in Ward 17? **5**

4 A shop has 50 boxes of polo shirts. Each box has 20 shirts. All the shirts are put out for sale. On the first day 637 are sold.

How many shirts are still for sale? **363**

5 There are 6 folders in each pack. A school buys 100 packs. The packs are shared equally between 10 classes.

6 folders

How many folders does each class get? **60**

6 A farmer has 124 bales in one field and 116 in another. The bales are collected and stacked in eights.

How many stacks of eight does the farmer make? **30**

7

When I am added to 7 and the answer multiplied by 100 followed by subtracting 450 you get 550.

What am I?

3

Review problems 6

1 Mel gets the 9:15 bus to the station. It takes 35 minutes. She gets on a train which should take 3 hr 20 min, but it arrives 5 minutes early.

How long was Mel's journey? **3 hr 50 min**

2 A market trader has 50 boxes of cabbages. Each box weighs 25 kg. She opens the boxes and puts the cabbages with the 13 kg she already has.

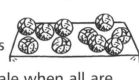

How many kilograms of cabbages has the trader for sale when all are together? **1263 kg**

3 Alec and Jane's aunt gives them £30 to be shared equally. Jane spends £6.99 of her money. Alec spends £8.55 of his.

How much has Jane left? **£8.01**
How much has Alec left? **£6.45**

4 Five men lay tarmacadam on three roads. Each road is 24 metres long. They run out of material 13 metres from the end of the last road.

How many metres of tarmacadam do they lay? **59 m**

5 Mr Watson harvests 60 acres of hay. Each acre produces 100 bales. His wagon can only take 400 bales each time to the store.

How many times does he fill his wagon so that all the bales are in the store? **15**

6 Mike has two rectangular plots for growing vegetables. One is 3 m by 5 m, the other is 4 m by 2 m.

What is the total area of his two vegetable plots? **23 m²**

7 In a cupboard John finds eight full bottles of lemonade. Each bottle holds 1.5 L. He and his friends drink 800 mL of the lemonade.

In L and mL how much lemonade is left? **11 L 200 mL**

(48)